BUFFALO

D0517302

Text © 2005 by Harold Picton
All rights reserved. No part of this work may be reproduced or used in any form by any means—graphic, electronic, or mechanical, including photocopying, recording, taping, or any information storage and retrieval system—without written permission of the publisher.
Printed in China
05 06 07 08 09 5 4 3 2 1
Library of Congress Cataloging-in-Publication Data
Picton, Harold D.
Buffalo : natural history & conservation / Harold Picton.
p. cm. — (WorldLife library)
Includes bibliographical references (p.71) and index.
ISBN 0-89658-727-4 (pbk. : alk. paper)
1. American bison—Juvenile literature. I. Title. II. World life library.
QL737.U53P533 2005
599.64'3'097—dc22
2005011156
Distributed in Canada by Raincoast Books, 9050 Shaughnessy Street, Vancouver, B.C. V6P 6E5
Published by Voyageur Press, Inc.
123 North Second Street, P.O. Box 338, Stillwater, MN 55082 U.S.A.
651-430-2210, fax 651-430-2211
books@voyageurpress.com   www.voyageurpress.com

Educators, fundraisers, premium and gift buyers, publicists, and marketing managers: Looking for creative products and new sales ideas? Voyageur Press books are available at special discounts when purchased in quantities, and special editions can be created to your specifications. For details contact the marketing department at 800-888-9653.

Map on page 49: 'Migrations of the Buffalo and their Hunters' from *The Buffalo* by Francis Haines.
Copyright © by the University of Oklahoma Pres, Norman. Redrawn with permission.

Photography copyright © 2005 by:

Front cover © Chris Knights / Ardea.com
Back cover © Jim Brandenburg / Minden Pictures
Page 1   © Peter Cairns / northshots.com
Page 3   © Peter Cairns / northshots.com
Page 4   © Bert Gildart / Still Pictures
Page 6   © François Gohier
Page 9   © Tim Fitzharris / Minden Pictures
Page 10 © Peter Cairns / northshots.com
Page 13 © Rod Planck / NHPA
Page 14 © Jim Brandenburg / Minden Pictures / FLPA
Page 15 © Thomas D Mangelson / Images of Nature
Page 17 © Mark Newman / FLPA
Page 18 © Tom & Pat Leeson / Ardea.com
Page 21 © Manfred Danegger / NHPA
Page 22 © Tim Fitzharris / Minden Pictures

Page 23 © David Hosking / FLPA
Page 24 © Peter Cairns / northshots.com
Page 25 © Parks Canada / Barret & MacKay / 09.91.10.01 (70)
Page 26 © M Newman / FLPA
Page 27 © T Kitchin & V Hurst / NHPA
Page 28 © François Gohier
Page 29 © D Robert & Lorri Franz
Page 30 © Parks Canada / Barret & MacKay / 09.91.10.01 (64)
Page 33 © D Robert & Lorri Franz
Page 34 © Peter Cairns / northshots.com
Page 35 © François Gohier
Page 37 © Erwin & Peggy Bauer
Page 38 © Peter Cairns / northshots.com

Page 41 © Erwin & Peggy Bauer / Bruce Coleman Inc
Page 42 © T Kitchin & V Hurst / NHPA
Page 48 © D Robert & Lorri Franz
Page 51 © Alamy
Page 52 © Kansas State Historical Society
Page 53 © Glen Bow Archives NA-2242-2
Page 55 © J Brandenburg / Minden Pictures / FLPA
Page 57 © D Robert & Lorri Franz
Page 58 © François Gohier
Page 61 © François Gohier
Page 62 © François Gohier
Page 63 © François Gohier / Ardea.com
Page 64 © Peter Cairns / northshots.com
Page 67 © David Hosking / FLPA
Page 72 © D Robert & Lorri Franz

BUFFALO

Harold Picton

WORLDLIFE
LIBRARY

Voyageur Press

To Merieon
Harold Picton

Contents

Introduction

I sat within a few yards of the coyote. We were both intently watching the herd of buffalo slowly grazing their way across the winter landscape. They swung their massive heads from side to side, clearing the snow from the buried grass. Steam blew from their nostrils into the frigid air, forming a small cloud of ice fog. Occasionally the coyote would dash near the bulls to grab a small mammal exposed by head swings of the buffalo. The temperature drop that came as the weak winter sun dropped below the horizon soon sent me back to my truck, but my wild companions continued their activities without pause.

The coyote may have been the same one that I saw in the late spring pursuing small mammals near a herd of buffalo cows and calves that I was watching. As the grassing lawn maintained by the buffalo was far too short for small mammals, the coyote walked through the herd without causing concern to either buffalo or the coyote as it proceeded to more productive hunting in patches of taller vegetation. The calves ran and played without a care, occasionally stopping to get a drink from their tolerant mothers. A cow dusted one side in a wallow, then stood, and dropped to do the same to her other side. A cow bird rode proudly by on the back of a shaggy steed, hoping to find an insect meal in the wads of hair being shed. Two young bulls were engaging in shoving matches on the edge of the herd. Occasional pivots and hooks made this a ballet dance in preparation for the coming reproductive season. All was well on this sunny day in this buffalo heaven.

My low-flying airplane skimmed over the horizon-to-horizon landscape of the Great Plains with its overwhelming arch of sky, but the large herds of wild buffalo are gone from their natal land. A ranch may have a hundred or two buffalo in a fenced pasture, but the great free-ranging herds are absent. Here and there amongst the farms and ranches one can still see the 'tepee rings' of native North American villages. Even traces of the Great North Trail used by some of the aboriginal settlers of North America can be found. Pronghorn antelope herds are once again abundant on this landscape. It is the buffalo, their historic grazing partners, which formed the core of the Great Plains ecosystem and of the life of the buffalo-hunting native North Americans, that are no longer to be seen. Therein is a tale of how this present circumstance came to be. It is a tale of the past, the present and perhaps the future. It is a tale worth the telling.

What's in a name – Buffalo or Bison, Bison or Bos?

Names have always been a source of confusion when discussing this species. When Europeans arrived in North America they used the words they already knew to describe the new animals that they encountered. A familiar name, buffalo, was used for the new animal. 'Buffalo' is also used for various unrelated African and Asiatic species that do not have a hump. For this reason 'bison' is more specific since it applies only to the North American animals. However, 'buffalo' is commonly used and is well established as a synonym for bison. In this book, 'buffalo' is used to refer to the North American bison.

There have also been discussions concerning names at the scientific level. Researchers that have emphasized the similarities between bison and cattle have proposed that they be grouped together in the genus *Bos*. This would make the scientific name *Bos bison*. Other, mainly North American, mammal specialists emphasize the differences between bison and cattle and have retained the name *Bison bison* for the species. Whatever we call it, this species is clearly an icon of North America.

Origins and Evolution

The largest land mammal of North America grazes slowly over the short grass prairie. Seldom called beautiful but rather tough, rugged, hulking or even bizarre, the American buffalo or bison (*Bison bison*) brings feelings of an ancient age. The bison lineage began 2 to 3 million years ago as a small member of the cattle family in southern Asia dwelling in forests and parklands. The earth was beginning to cool and ice was going to move south over the land. This Pleistocene era was a time of climate change and that change brought opportunity. The genetic line that began in southern Asia seized the opportunity to invade the vast grasslands to the north. With increases in size and modifications of form, to match the new environment, the steppe bison (*Bison priscus*) emerged to carry the genetic line across Eurasia to the Bering Straits.

Between 187,000 and 129,000 years ago the Illinoian Ice Age provided another new opportunity. This is when sea levels lowered to form a land bridge called Beringia between Siberia and Alaska. This allowed steppe bison to amble into the New World to begin the saga of the North American bison. The most recent common ancestor of both the Eurasian and North American bison seems to have existed about 136,000 years before present. As the ice of the Illinoian Ice Age vanished, sea levels rose, severing the connection to Asia. A warm period of 50,000 years allowed buffalo to extend their range into the central grasslands of North America. Genetic information from fossil DNA suggests that the buffalo population was increasing steadily.

About 70,000 years ago, the warmth again faded and the Wisconsin Ice Age began, reaching its maximum extent 22,000 to 18,000 years ago. The great ice sheets built slowly across the continent from the area of Hudson's Bay. This left a band of grassland, east of the Rocky Mountains, extending from Central America across Beringia to Siberia.

The first 35,000 years of this period created a buffalo paradise. Beringia had re-emerged from the sea. Analysis of fossil DNA suggests that there was a two-way exchange of buffalo between Asia and North America. They now ranged across Eurasia south to Central America. This vast area provided the buffalo paradise from 60,000 to 37,000 years ago. The population appears to have been doubling every 10,000 years and may have reached 200 times the population that was present when the west was settled. This was the peak in genetic diversity as well as in numbers.

A representative of this buffalo golden age is present in the form of a frozen fossil of a steppe bison named 'Blue Babe.' It has been given a radiocarbon date of about 36,000 years before present. Blue Babe is currently located in the museum at the University of Alaska in Fairbanks. This adult male is estimated to have weighed about 1600 to 1700 lb (700 to 800 kg). An American lion apparently killed it.

As the years spun past, after the 37,000 years before present mark, increasing cold brought fragmentation and loss of habitat. A general population decline began that continued virtually unbroken through the 19th century. The climate change brought spruce forests intruding in to the plains habitat essential to buffalo. A landscape of ice then expanded to isolate the population of the southern plains from that of Alaska and the Yukon. It appears that the isolation may have been virtually total from 22,000 to 12,000 years before present. This

This skull shows the shock-absorbing struts that connected the inner brain case and outer skull, protecting against concussions in head-to-head combat.

time of isolation allowed the various buffalo populations to go their own ways. Several different body types appear in the fossil record. The relationships between these various types are not entirely clear, although genetic analysis suggests that the southern or Great Plains population contributed most to the diversity. This period when the ice melted was a time of great change. It was the period of the great Pleistocene extinction in which over half of the large

mammal species then living disappeared forever. The buffalo is one of the survivors.

The buffalo in what is now the continental United States were partially or totally isolated from the northern populations for at least 20,000 years. As they went their own way, they developed into the giant or large horned buffalo (*Bison latifrons*). This animal had a horn spread of over 7 ft (2.15 m) in comparison with the 3 ft (1 m) horn spread of today's buffalo. It is estimated that the weight of the giant buffalo was twice the weight of the modern form.

Under the impact of climate change it is thought that the northern steppe bison gradually developed into a newer edition called *Bison occidentalis*. Fossils suggest that, as the ice retreated, some northern bison had occupied some areas of northern Alberta by about 10,000 years ago. DNA from fossils also indicates bison from the southern population had colonized this far north before the spruce forest of the taiga closed in. The northern and southern populations may have come into contact during this time.

The studies of DNA from fossils and living buffalo indicate that all living forms of North American bison are offspring of the Great Plains population. It is clear that the bison numbers and genetic diversity continued to decline between 20,000 and 10,000 years before present.

Although the ice had gone, the giant bison and its successor, the ancient bison (*Bison antiquus*), faced yet another challenge from the climate. Another period of extreme climate, called the Hypsithermal, provided about 4000 years of challenge. Resulting from the regular changes in the tilt of the earth's axis, this period  brought hotter summers and colder winters. It brought extended droughts to the Great Plains lying east of the Rocky Mountains. Strong winds tore dust from the dry beds of the many very large glacial lakes. The great vulnerability of the North American continent to climate change and its propensity to produce extinction of animals forced bison to adapt or die.

In the 4000 years or so of this period and in a thousand generations the ancient buffalo became smaller and developed the characteristics we see today. By 5000 years before present, buffalo, as we know them, occupied the continent but with a smaller genetic pool than was present in earlier eras. The adaptations developed during the Hypsithermal period worked and the buffalo began to increase their numbers.

While it is thought that climate and its effect upon habitat was the primary factor producing these adaptations, other significant changes were also taking place. The giant-sized predators of the ice age, such as the American lion and the giant short-faced bear, vanished. Grassland competitors such as native horses had also departed by 8000 years before present. A new element, man the hunter, entered the picture. Man, in the form of the Clovis culture (the earliest definitively dated human populations in the Americas), was certainly present by 13,000 years ago. But we also should note that artifacts at some sites such as the 'Old Crow Site' in the northern Yukon might indicate that a few people were present even 42,000 years ago.

Today, we recognize two forms of buffalo, separated by the taiga that stretches across Canada. Originally native to Alaska and northwestern Canada, the wood bison (*Bison bison athabascae*) is the larger of the two subspecies. The plains buffalo (*Bison bison bison*) occupied the Great Plains from southern Canada to Mexico. Millions of buffalo and pronghorn antelope prospered in a grazing system that filled the short grass prairie east of the Rocky Mountains. Its distribution also extended into forest parklands east of the Mississippi River to the eastern slope of the Allegheny Mountains and into northern Florida.

The bison of Asia disappeared about 2000 years ago but the bison heritage has been continued in the Wisent or European bison (*Bison bonasus*) that still lives in the forests of eastern Europe.

A wood buffalo in Elk Island National Park, Alberta, Canada.

Wind and dust are frequent companions on the Great Plains.
Note that the calf does not yet have the fully developed shoulder
hump of its mother. Custer State Park, South Dakota.

Form and Function

The beauty of the buffalo is seen in the physical and behavioral adaptations that developed in response to climate, food supply, predators, their own social life, and the impact of man-the-hunter for 10,000 years or more.

Food and feeding are important in shaping an animal. Buffalo belong to the group of mammals that assist their digestive processes by regurgitating their food and chewing it a second time. This is 'chewing their cud' or ruminating. Thus they are called ruminants. Their digestive specialization includes a four-part 'stomach'. Two of these chambers (recticulum and rumen), actually part of the esophagus, form a large fermenting vat. Because ruminants can't digest cellulose they hire bacteria to do it. The bacteria are paid by taking some of the energy from the cellulose in the plants gathered while leaving most of the energy for the host animal. The host animal also pays the rent and utilities in providing a constant environment and by gathering food. Bacteria break down the abundant cellulose in the plant material that has been eaten. This process produces some chemicals that the host can use. The bacteria doing this job are produced in enormous quantities, and this bacterial 'yogurt' is then passed into the two chambers (omasum and abomasum) of the true stomach to be digested. The bigger the animal, the bigger the fermenting vat, the longer the food can be 'cooked' and the lower the food quality of the plants can be. A small ruminant, like the ancestral forest buffalo, can only have a small rumen because they have so many predators that want to eat them. After all, one can't run very fast with gallons of fluid sloshing around with each step. Because small ruminants can't hold much food they have to select only the most nutritious bites. A narrow lower jaw, functioning like tweezers, enables them to be very selective in feeding.

The ice age buffalo lived in a different environment, the 'mammoth steppe'. The relatively tall plants of this vanished grassland type provided food for the bison as well as the mammoth. While the buffalo's jaw was wider, it was still a more selective instrument than we see in today's buffalo, which in contrast have a broad lower jaw with a relatively straight row of incisors. This is a 'cutter bar' that is optimized for gathering large amounts of low-growing vegetation. A large fermenting vat that can 'cook' a lot of low-quality food is coupled with this jaw. The jaw is in a massive head that hangs close to the ground, bringing the teeth close to the short grass. Huge muscles are needed to support the large head. These muscles are anchored to neural spines, about 7 in (18 cm) in length, that rise from the vertebrae of the back. The neural spines and the attached muscles form the shoulder hump. The food-gathering mechanisms and body form thus reflect the intense battle to survive and the competition for food that the change from the ice-age climate brought about.

The forward position of the hump seen in living buffalo permits the quick pivot and disemboweling hooking tactic used to deal with wolf packs. Ice-age buffalo had to contend with predators that were larger than those of today, such as the 800 lb (364 kg) American lion and 2000 lb (910 kg) giant short-faced bear. The hump on the steppe bison was located further back, which improved the efficiency of running for long distances to escape the giant predators. Even with their forward hump, modern buffalo are efficient runners, particularly for shorter distances, and can attain speeds of 36 miles (60 km) per hour.

The intense struggle for survival during the last 30,000 years brought an intense competition to reproduce. The male steppe buffalo and their predecessors engaged in display rituals and brutal shoving matches to determine whom were to father calves. Their long horns tended to prevent opponents from

slipping by and gouging their sides. The climate then became hotter and drier. The body of the buffalo became smaller, which provides a greater proportion of surface area to help lose heat. The smaller body size also became more physiologically desirable to meet the decreased food supply. At the same time the reproductive combats became more violent.

To protect against brain concussions the front of the skull became domed, with an outer layer of bone and shock-absorbing struts which connected this outer skull with the bone of the brain case. A bonnet or pad of thick, shock-absorbing hair developed on the head. The horns moved back on the skull where they would not be damaged by collisions. They also shortened and developed more of a hook. Horns serve defense as well as offense. The hook gives the ability to lock the horns of the opponent when on defense. If the opponent is weaker and allows its body to twist, a quick upward slash can inflict serious injury to the opponent's side.

The forward position of the hump, which allows an exceedingly rapid pivot, coupled with a disemboweling hooking ability, makes them into a truly fierce opponent for either predators or other buffalo. The thick hair coat on the fore part of the body serves as armor against hooking and pivoting attacks as well as against the slashes of predators. The battles during the rut are not entirely 'show' or ritual battles. They leave injured and dying bulls each year.

Characteristics

The two subspecies of buffalo differ in some respects. The bulls of the wood buffalo may weigh 2000 lb (910 kg) and adult cows 1247 lb (567 kg). Plains buffalo are smaller, with adult bulls reaching 1691 lb (769 kg) and females 1000 lb (454 kg). The plains buffalo living on Santa Catalina Island were transplanted from Yellowstone Park but now average about 10 percent smaller. Wood buffalo are usually darker and lack the 'chaps' or 'pantaloons' of long hair on the upper forelegs seen on the plains buffalo. The peak of the hump in wood buffalo lies in front of the shoulder, while the high point of the more rounded hump of the plains buffalo is above the shoulder. The tail of wood buffalo is longer and more heavily haired. The hair on the forehead is longer on the wood buffalo but the plains animal has a bigger beard.

Wood buffalo horns extend above the hair of the head.

Wood buffalo have a hump that has sharper contours both in front and back than does the plains buffalo. The neck of the plains buffalo is shorter than that of its relative.

Buffalo are remarkable grazing machines. Grasses and grass-like sedges make up the bulk of their diet. Although classed as roughage or bulk feeders, buffalo are quite selective in the plants that they feed upon. Roughage feeders are grazing animals that focus more upon quantity than upon quality in selecting their food plants. Buffalo will voluntarily graze on steeper slopes than will cattle. Feeding will take 9 to 11 hours each day on a range of good quality. Feeding in the morning and again in the evening is typical but they will feed even at night in some circumstances. They seem to process low-quality foods more efficiently than do cattle. When grazing, the large cow and calf herds break up into groups of 10 or 20 and spread out over the area to maximize access to ungrazed vegetation.

Plant foods naturally contain tiny particles of silicon dioxide (glass or opal) that are very abrasive. These particles wear down the teeth of animals feeding upon them. Old age in buffalo means worn-out teeth and starvation. Animals that have reached an age of 15 years are considered to be old, although some have lived in captivity to 41 years.

The manure deposited by the buffalo serves as a nitrogen fertilizer for the plants, which makes the plants more nutritious and brings the grazing bison back to the same area at a later date. This leads to the creation of areas of short clipped grass called 'grazing lawns.' In some areas prairie dogs occupy and maintain grazing lawns in a cooperative relationship. Since other animals such as burrowing owls, come in and also use the prairie dog towns, bison are sometimes regarded as a 'keystone species', whose presence is key to maintaining other species in the ecological system.

Buffalo have produced grazing lawns amidst the sage brush here in the Lamar Valley of Yellowstone National Park, Wyoming. This terrain forms part of the 'Buffalo Ranch' that was so important in the restoration of buffalo to Yellowstone Park.

Family Matters

A few buffalo of both sexes become sexually mature at about one-and-a-half years of age. Most become mature a year later. A high percentage of two-and-a-half-year-old cows become pregnant. As is common in wild species, the mortality of the calves of first-time mothers is considerably higher than that of mothers with more experience. The ages from four to ten represent the prime breeding years for cow buffalo.

Males younger than six are not socially mature and have little chance of participating in the breeding sweepstakes. Bulls that are eight to ten years old have their best chances to become fathers. While most cows become mothers many bulls do not succeed in becoming fathers.

Adult bulls spend most of the year alone or in small groups separate from the cow-calf groups. They become more social as the summer progresses. The breeding season is the big social event of the year. August is usually the peak for romantic activity but it does vary geographically. The bulls bellow challenges at each other. Both the bulls and cows dust themselves more frequently. Wallowing in urine-soaked dirt forms a 'beauty potion' for the rut. This seems to be part of the mechanism that encourages the cows to come into estrus and helps to coordinate the breeding cycle of the herd. The mature bulls keep a close watch on the cows to determine those coming into breeding condition. They do this by 'flehming' or 'lip curling'. This exposes some special sensory receptors in the upper gum area that are part of a second odor detection system specialized for detecting reproductive odors. The bulls 'tend' or stay with the cows that show breeding readiness. The cows do have a say in the matter. They will usually not let a bull of low rank stay near them. Most of the actual breeding is at night.

Wood buffalo have shorter beards and less forehead hair.

Bulls escorting a cow are frequently challenged by other bulls. To assess their opponent, the bulls enter into a 'parallel walk' as they challenge each other. If neither backs down, they may charge, with two tons of buffalo colliding head to head. A quick pivot may then be used to gore the opponent in the side. Most challenges are decided without injury in the parallel walk stage or with a head-to-head shoving match. The competition does go farther, with injured and exhausted bulls becoming more common as the rut proceeds.

A number of the injured bulls that I have seen have had puncture wounds, large gashes or some other injury to their hips. Similar injuries to the abdomen or those that break bone can be sentence of death. One study found that 50 percent of the bulls showed evidence of previously fractured ribs. Bulls that die during the rut form a major source of food for coyotes, grizzly and black bears, wolves, ravens and other scavengers in areas such as Yellowstone National Park.

Wood buffalo live in small groups using openings in a forested landscape. Their breeding system differs, probably because of the social and habitat differences. Lone wood buffalo bulls gather harems of cows, which they tend. They will attempt to defend their harems against other bulls that happen along. Because the breeding peak of a given bull may last only a week or two, the harems will probably have several different males with them during different periods of the rut.

New life in the herds begins with the birth of single calves at the end of 262 to 300 days of pregnancy. Twins are very rare, occurring less than once in several thousand births. Most births occur in a four-week period in April and May, although one occasionally sees calves that were born a month or two later. Such small, late-born calves are unlikely to survive the rigors of winter. Birthing is a social process in herds that occupy open grasslands. Other cows come and sniff and lick the newborns. This neighborly care is the route for the transmission of brucellosis if the calf is stillborn due to the disease.

The reddish-colored calves can stand and suckle shortly after birth. They differ in appearance from the adults so much that people have been known to ask, 'Why are those big red dogs following the buffalo around?' The calves lack a hump. Like most ruminants, their digestive system is specialized to handle milk at birth, not vegetation. They are unable to digest plant material until they are about a month old. The 'fermenting vat' portions of the stomach chambers develop during this first month. Their mothers provide the bacteria needed to get their 'fermenting vats' going by licking their faces. They learn which plants to eat by watching their mothers while in 'food habit school'; many plants are unsuitable as food and some are downright poisonous.

Buffalo are highly social and calves are watched over by all adults of the herd. Energetic play rules the day with well-nourished calves, but after harsh winters or droughts many calves are listless because of insufficient milk from their malnourished mothers. Play serves not only the function of physical conditioning but also enables the young to learn the social rules of the herd. Watchful wolves and bears prey on calves straying out of the equally watchful herd. If wolves are present the calves are often gathered into the front and interior portion of the herd. In the historical days of large herds, it was noted that a mother would defend her calf when beset by wolves until the rest of the herd moved on enough to make her vulnerable. At that point she would abandon the calf. This satisfies a basic rule of natural economy that says that it is best to cut one's losses. It is better to sacrifice the calf, in which there is only a few months of energy invested, than to sacrifice a mother that can breed again and which has years of energy invested in it. Energy that has been captured in usable form is the currency of the natural system. In a recent incident, a female grizzly bear chased a buffalo cow with a calf. The bear downed the calf with a blow of her paw and then began eating it. The buffalo cow stayed near, watching and threatening, for about 30 minutes before leaving.

Calves separated from their mothers can be very tame. Native North Americans, settlers and many others have kept them as members of their households. This tranquil picture changes as the buffalo become older. The animals become more unpredictable and aggressive as they mature. A number of people have been killed by seemingly docile pet buffalo. People often underestimate the agility of buffalo. In one case, a tourist seeking to take a picture was gored, thrown in the air and came down on the buffalo's back where the animal quickly turned its head to gore the person a second time.

Getting lunch from mother. Note the reddish color and absence of a hump of these two- to three-month-old calves. The manure from the grazing animals fertilizes the grazing lawn, which makes it more nutritious and encourages repeated grazing.

Habits and Habitats

As herd animals, buffalo communicate with each other vocally, with body language and with odors. They are most vocal during the rut when grunts, bellows and roars are common. While bulls are the noisiest, cows join in the din. Much of the communication involves social dominance within the herd. While size and age are involved in the dominance relationships, they are not the only factors. Males tend to dominate females. The social position of the calf reflects that of its mother.

In addition to its role as a fly swatter, the tail also serves as a signal flag. A tail up in a fight between bulls usually indicates dominance and tail wagging, submission. A tail up by a cow may mean there is a predator in the vicinity.

As the new summer coat grows, rubbing away the old coat becomes a prime activity. Virtually any rough surface will do. Rubbing is a forceful activity, which may badly damage small groves of trees. Telephone and power poles are appreciated, particularly if they have been studded with nails. Both sexes also use small trees and shrubs that can 'fight back a little'. These are used to practice their horning and upward-thrust goring techniques. A buffalo rubbing against a corner of a cabin, while one is sleeping in it, will certainly get your attention.

Wallows are used for dusting. Buffalo are unable to roll over because of their hump. They first lie on one side, then stand and lie on the other side. To make wallows, vegetation is usually scraped from an area with their hooves, leaving a shallow pit of dry dirt. They will sometimes use these when wet and muddy. All sexes and ages enjoy a good romp in the dirt. The dusting activity is believed to help control biting insects and parasites. Wallowing activity increases as the rut develops.

Buffalo swim well. Just the nostrils and tops of their heads ride above the water. It sometimes seems as if adults maneuver to get upstream and break the current for a calf swimming with them in rivers. Many travelers noted large numbers of drowned buffalo along the Missouri and other large rivers. They had broken through or gotten trapped by ice on the rivers or died while trying to swim flooded rivers. Three separate drowning incidents killed 500, 1100 and 3000 wood bison during floods of the Peace River in Canada in the 1950s and 60s.

When storms move across the open plains, bison stand facing into the storm while cattle turn their tails. The heavy wool coat on the head and chest provides protection against the storm. Their ice age ancestral lineage enables them to handle winter well. They swing their massive heads to clear the snow from forage. Bison can handle snow depths of up to about 18 in (46 cm). Their ability to use low-quality forage helps them to survive the harsh winter conditions.

Summer heat is handled by bedding during the heat of the day, using the insulation of their coat and a 'thermal lag' strategy to prevent overheating. This strategy simply reflects the fact that the insulation on the back delays the transfer of solar heat to the body, and slows the warming of the body until the heat of the day is past. Staying bedded reduces the heat received by reflection from the surrounding ground. At the warmest times of the day, the surface of the ground is often cooler than the air. The body of a bedded animal is in contact with this cooler surface. Becoming more active as the day cools, they stand and expose the bare skin of their 'thermal windows' to get rid of excess heat. Smaller body size gives more body surface area for a given amount of

A bull plains buffalo getting a good dusting. It cannot roll over, due to its hump. It will have to get up and lie down on its other side. Dusting is believed to help with control of insects and skin parasites.

Unlike cattle, buffalo face into the storm rather than away from it.
The snow depth in this scene from Yellowstone National Park is about at the
maximum for its easy movement through it.

body volume. Because of this, one route open to species that developed in cold environments is to become smaller in warm climates. This is seen in bison where the southern plains bison were smaller than those of the northern plains. The bison of Santa Catalina Island of southern California are smaller than their relatives living in the cold climate and high altitude of Yellowstone National Park. Wood bison are also larger than plains bison.

Buffalo have long been noted for their ability to select the most practical travel routes. Their trails were often used as the best travel routes for man in the early days of the settlement of the west, and many highways and railroads follow routes originally pioneered by buffalo. Movements of 48 miles (80 km) in a day have been measured. Free-ranging plains buffalo cows have been reported to have a home range of 196 square miles (540 square km) in Yellowstone Park. Some wood bison cows may have double this size. Home range sizes vary by sex, age and the type of habitat involved.

Winter travel is little affected by moderate snow depths. The buffalo maintain single-file trails in deeper snow. Mature and wary cows often lead the single files. The 'tail-enders' are vulnerable to ambush by wolves lying in wait. Buffalo will use plowed highways and packed snowmobile routes but they show no preference in using them over their own trails through the snow, which often parallel the human routes. There are some things that one doesn't really know how to interpret. During an aerial survey in early spring we once circled to watch a big bull buffalo that came to a large snowdrift in a small gulch. It started through and found the snow too deep and backed out. It then charged the drift and made it a little further, backed out, and took another run. It finally broke through on the fifth charge. The strange thing was that it could have walked 30 yd to either side to pass the drift on easily visible bare ground.

Their movements can become truly awesome. A stampede may result if an already wary group of buffalo is disturbed. In 1882 a group of hunters were in their tents when a low rumble was heard which grew louder and louder. A huge herd of buffalo was stampeding toward them. The men ran out of the tents yelling, banging pans and firing guns to induce the herd to split and keep it from pounding the men and camp into the dust.

In the days of using buffalo jumps, native American hunters developed a hazardous but effective method for guiding stampedes. A horseman draped in a buffalo hide would get in front of the herd, riding moderately rapidly in the direction they wished the buffalo to take. Others would then trigger the stampede and the bison would follow the decoy. The rider had to be very agile and skilled to avoid being driven over the cliff or trampled by the stampeding herd. The same method was used before the days of horses by having the fastest runner in the tribe play the role of decoy.

Not many predators view live buffalo as food. Grizzly bears are known to kill a few cows and calves. During an aerial survey, my pilot and I once watched a grizzly chase an elk through a patch of timber. Coming out of the timber the bear nearly ran into a bull buffalo. The bear showed due deference to the buffalo, which was five times as heavy as the bear. The elk got away. In another case, a female grizzly bear attacked an unwary three-year-old bull buffalo, crippling it with one blow. The buffalo was probably over twice the weight of the bear.

Predators in a given area often specialize in hunting a single species. The wolves reintroduced into Yellowstone National Park have been specialists in hunting elk. Some wolf packs have now begun to hunt buffalo, as the elk population declined and the buffalo expanded. Wolves have accounted for as much as 30 percent of the mortality in Canada's Slave River Lowlands buffalo population.

*The thunderous ground-shaking hoof beats of a buffalo stampede brought
fear to many travelers of the Great Plains. Riding in amongst a herd such as this
was an exciting method of hunting used by native North Americans.*

Humans and Buffalo

Man is certain to have entered the picture in North America by 13,000 years ago. These newcomers quickly took advantage of the gourmet food supply offered by the buffalo. Humans developed an intimate connection to buffalo over these thousands of years. Knowledge of this past is necessary if we are to understand the current status of buffalo and to propose conservation strategies for the future.

The people of the Clovis and later cultures were effective hunters and found the buffalo a rich food supply. This was the period of rapid buffalo evolution and the hunters influenced people and were influenced by them. The hunters from this time forward passed on the knowledge necessary to conduct hunts through songs, myths, dances, rituals and codes of behavior. The portions of the legends we read are but the simplest verses from those told around the dancing firelight in winter lodges. The stories give a view of nature, the laws, the customs and knowledge about how to carry out the hunt.

Archeologists have found hundreds of kill sites going back over 11,000 years on the Great Plains. Most of these are in small valleys or other sites where traps could be constructed and small groups of bison driven in and killed. Traditional knowledge says that in these circumstances all of the buffalo in a small group must be killed or 'they will go back and warn the rest of the buffalo'. Such precautions were apparently necessary. Alexander Henry of the North West Fur Company tells of a situation where buffalo were consistently hunted individually for food. The buffalo became very wary and fled at the barest smell or sign of a human.

One of the verses from an epic tells of a starving group of people who appeal to the supernatural Sun Buffalo Cow for aid. Sun Buffalo Cow felt their need. Saying she would help them, she changed into an earthly buffalo. She then dove to her death over a cliff, feeding her people and insuring buffalo to hunt.

Hunts were also conducted by driving buffalo over a small cliff or 'pishkin'. Some of these 'buffalo jumps' or 'pishkins' were used for generations. It is estimated that over 10,000 buffalo are represented in the bone beds, accumulated over 300 to 500 years, at buffalo jumps such as the Ulm site in Montana and the Vore site in Wyoming. The Highwood site in Montana contains over 200,000 buffalo remains. A number of these bone beds were mined as fertilizer in the late 19th century. The crushed body of a boy found at the 'Head-Smashed-In' site in southern Alberta, gives testimony to the dangers involved in this method of hunting.

The Lakota began their intimate relation with the buffalo when a beautiful young woman, changing into a white buffalo calf, arrived at their village. She gave the tribe the rules by which they were to live. In leaving, she changed first into a brown buffalo and then back into White Buffalo Calf Woman and vanished. The herds of buffalo to hunt were her legacy. The rare occurrence of a white buffalo reminds the tribes of her commitment to them.

In 1804 under orders from President Jefferson, Captain Lewis and Captain Clark and the Corps of Discovery explored a route across the continent to the Pacific Ocean. Buffalo provided the expedition with a surplus of food while they were east of the Rocky Mountains. While spending the winter of 1804-05 with the Mandan Indians, Captain Lewis noted that buffalo wool is very soft and considered its potential for cloth.

The journals of Captain Meriweather Lewis provide us with a view of a migrating herd that he encountered along the Missouri River of Montana on July 11, 1806. 'I arrived in sight of the white bear islands – the Missouri bottoms on both sides of the river were crouded (*sic*) with buffaloes – I sincerely believe that there were not less than 10 thousand buffaloe (*sic*) within a circle of two miles around that place.'

The best estimates that have been made suggest that perhaps 25 to 30 million buffalo ranged over the Great Plains, with another 10 million elsewhere on the continent. Those east of the Mississippi River disappeared rapidly as Euro-American settlement progressed. The wood buffalo herds in Alaska and the neighboring Yukon vanished by the mid 1800s. Hunting by native hunters may have brought about this extinction as they struggled for survival during the severe weather of the Little Ice Age (1350 to 1850). This reduced the wood buffalo range to the area south of Great Slave Lake in northwestern Canada. The scattered and small herds of bison west of the Rocky Mountains on the plains of the Columbia River and eastern Oregon were gone by 1800. The few bison that occupied Idaho's grasslands were eliminated in the early decades of the 19th century.

The complex story of the death of the enormous herds that ranged over the Great Plains is focused upon the 19th century, although it began earlier. While genetic evidence suggests buffalo populations had been declining for thousands of years, tens of millions of them still remained on the short grassed plains in the 19th century.

The horses originally native to North America had disappeared by about 8000 years ago. The Spanish brought horses back to the Americas in the early 1500s. The horses spread northward, reaching the Canadian border by the middle of the 18th century. Some native American tribes recognized their utility and quickly developed the remarkable horse-buffalo hunting cultures. As the horses multiplied, they also competed with buffalo for food. An estimated two million horses impacted the southern buffalo herd by the early decades of the 19th century. Feral cattle in Mexico and southern Texas also increased the competition. About 1800, the livestock-borne diseases of anthrax and bovine tuberculosis entered the buffalo herds from Louisiana. The epidemics that raged through herds that had never been exposed to the diseases took their dreadful toll. This legacy continues today in some of Canada's wood buffalo herds. Brucellosis was added to the scourges in the late 1800s. Each of these diseases has different effects: anthrax kills relatively rapidly, bovine tuberculosis kills slowly and brucellosis causes abortion of the first pregnancy in buffalo.

Humans across the world were attempting to keep warm in the Little Ice Age, which waned in the late 19th century. Furs and native North American-tanned buffalo robes became trade items of high value. These native buffalo robes were the focus of trade of the American Fur Company and other companies. The thinner, more easily tanned hides from two- to five-year-old buffalo cows were preferred for the robes. These animals were also more preferred for food, except during the spring months when they were in poor condition. Skilled native American hunters took tens of thousands of these animals just before they entered their prime breeding years. Although this lowered the growth rate of the herds, their numbers still seemed inexhaustible.

The Great Plains is not an easy environment in which to thrive. Natural hazards were abundant. Many buffalo drowned crossing flooded rivers or after breaking through the ice in winter. Predation also occurred, as Captain Clark noted on October 20, 1804:

'I observe near all gangues of Buffalow (*sic*), wolves and when the buffalow move those animals follow, and feed on those that are too pore (*sic*) or fat to keep up with the gangue.' Frequent droughts and patchy precipitation

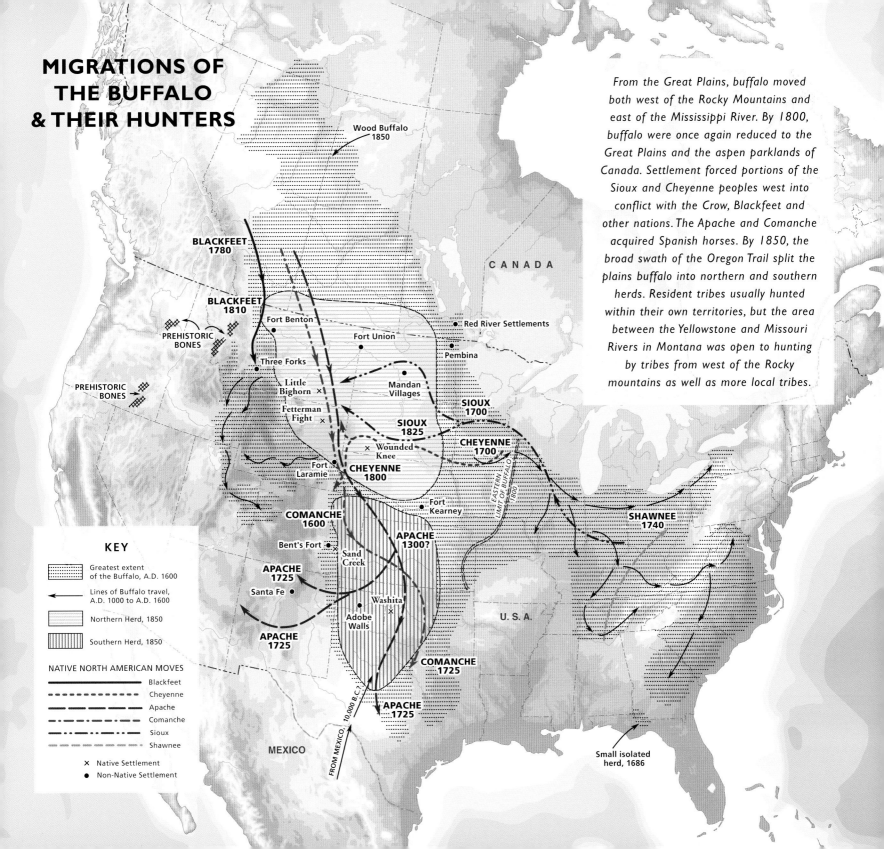

MIGRATIONS OF THE BUFFALO & THEIR HUNTERS

From the Great Plains, buffalo moved both west of the Rocky Mountains and east of the Mississippi River. By 1800, buffalo were once again reduced to the Great Plains and the aspen parklands of Canada. Settlement forced portions of the Sioux and Cheyenne peoples west into conflict with the Crow, Blackfeet and other nations. The Apache and Comanche acquired Spanish horses. By 1850, the broad swath of the Oregon Trail split the plains buffalo into northern and southern herds. Resident tribes usually hunted within their own territories, but the area between the Yellowstone and Missouri Rivers in Montana was open to hunting by tribes from west of the Rocky mountains as well as more local tribes.

Wood Buffalo 1850

CANADA

Red River Settlements

Fort Benton

Fort Union

Pembina

BLACKFEET 1780

BLACKFEET 1810

PREHISTORIC BONES

PREHISTORIC BONES

Three Forks

Little Bighorn ×

Mandan Villages

Fetterman Fight ×

SIOUX 1700

SIOUX 1825

CHEYENNE 1700

× Wounded Knee

CHEYENNE 1800

Fort Laramie

COMANCHE 1600

APACHE 1300?

Fort Kearney

EASTERN LIMIT OF BUFFALO 1800

SHAWNEE 1740

Bent's Fort

Sand Creek ×

APACHE 1725

Santa Fe

Washita ×

Adobe Walls

U.S.A.

COMANCHE 1725

APACHE 1725

APACHE 1725

MEXICO

FROM MEXICO 10,000 B.C.

Small isolated herd, 1686

KEY

Greatest extent of the Buffalo, A.D. 1600

Lines of Buffalo travel, A.D. 1000 to A.D. 1600

Northern Herd, 1850

Southern Herd, 1850

NATIVE NORTH AMERICAN MOVES

Blackfeet
Cheyenne
Apache
Comanche
Sioux
Shawnee

× Native Settlement
● Non-Native Settlement

*A cow and calf buffalo herd emerges from the Yellowstone River
in the morning fog. The Yellowstone Park herd is the largest and most
genetically diverse of all of the conservation buffalo herds.*

encouraged buffalo to use migration as a way to prosper. In the central and northern portions of the plains, great clouds of billions of Rocky Mountain locusts swept down from mountain valleys. About once every three years, these voracious insects would strip the vegetation from areas sometimes as large as 190,000 square miles (492,000 sq km). Grazing animals in these belts had to move or starve. The Rocky Mountain locusts became extinct at the end of the 19th century. The massive migrations born of this variable environment led hunters to believe that the supply of buffalo was without end.

Before horses arrived, humans traveled on foot using dogs as pack animals. The migratory behavior developed by buffalo also served to protect them from humans with their limited mobility. Natives burned rangelands at certain times of the year to attract buffalo to the new green grass near their villages. Sometimes the fires escaped to add yet another bit of variability to the environment. The arrival of horses changed this 'chess match' between humans and buffalo that had been going on for thousands of years. The behavioral strategies developed over the centuries worked well enough to allow the herds to prosper when hunted by people on foot. They still worked marginally well even after horses enabled humans to follow the herds and range more widely to find them. The hunters on horseback had to ride in close to kill with bows and arrows. The pivot and gore defense made this type of hunting thrilling but also very dangerous for horses and riders. It was the ability of the gun to kill at a distance that defeated these behavioral adaptations.

As the 19th century unfolded, the livestock herds with their diseases pushed north. The great movement of settlers to the West Coast began, and their herds of livestock stripped the grasses from a miles-wide strip along the Oregon Trail. In the 1860s, the invention of a new tanning technique, which made it possible to tan the heavy hides of older adults and bulls, began the last chapter for the buffalo herds. The hides became the prime material for the heavy belts driving the machines of the industrial revolution sweeping across Europe and America. Before this, wildlife including buffalo had provided food for the native North American residents of the plains and the increasing numbers of settlers, soldiers, gold seekers and railroad construction crews. Now the buffalo was firmly linked to the industrial economy of Europe and America.

In the 1870s, as the southern herds became depleted, the U.S. Army of the Montana District gave safe passage to native hunting parties to help tribes hunt buffalo on the plains of eastern Montana. This policy was maintained even through the annihilation of the Custer command, and the Sioux and Nez Perce wars. It was developed largely because the U.S. Congress did not provide money to feed native North Americans confined to reservations. The 600 to 3000 army troopers were too few to patrol the over 150,000 square miles (414,000 sq. km) of mountains and plains involved. The command elected to reduce native North American and settler conflicts by providing safe passage escorts past the settlements. Although somewhat haphazard, the civilian permit and military escort policy did allow the traditional hunt for food to go on until the herds vanished. Some native hunting parties made a 1000 mile (600 km) round trip annually from the slopes of the Cascade Mountains in western Washington. Each of these tribal hunts took hundreds to a few thousand buffalo. Meat was sliced thin and dried and then loaded on horses with the hides and other useful materials for the return trip to their home territories. The buffalo hunt meant more than food to the hunting tribes; it meant real wealth in every sense of the word. All parts of the animal were used and trade materials were made. The buffalo even produced the fuel to cook the meat

*A peacefully moving buffalo herd. Buffalo are active animals that cover several
miles in a day while feeding or simply moving to a new area of rangeland. Migrations produce
broad 'buffalo roads' that are famed for taking the best route between destinations.*

with, in the form of dried manure or 'buffalo chips', which provided a smokeless, odorless fuel for the peoples of the treeless plains.

The settlers, railroad construction workers and soldiers as well as natives hunted buffalo and other wildlife for food. The demand for wild game as food would probably have been enough to deplete the herds over time. However, it was the hunting for hides for the industrial market that brought the herds to a

rapid end. The southern herds were depleted by 1874 and the northern herd then became the focus of the hunting effort.

The hide-hunting teams consisted of a marksman with one to six skinners. Hides sold for $2.25 to $3.50 each. Depending upon where they bought supplies and equipment, a team needed to sell from 500 to 3000 hides to cover the basic expenses. Although killing an individual buffalo was relatively easy, it took considerable skill to be able to kill 25 or more in a small area. Marksmanship was a key, not only because a hide would pay for only ten or eleven cartridges, but also to avoid alarming the buffalo. Most carcasses were left to rot. A few small unsuccessful enterprises to save and market some of the meat involved digging pits in the prairie and lining them with 'green' or fresh hides to serve as pickling vats for choice cuts. The wastage of the meat and decline of the herds brought much concern from many people at the time. It has

In this bone yard in Michigan (above) the buffalo bones were ground into fertilizer.
Some 40,000 hides lie in this 1878 buffalo hide yard (left) in Dodge City, Kansas. Hunting for buffalo hides
for the industrial market brought the buffalo herds to a rapid end. By 1883, the great hunt was over.

continued to bring dismay to generations since. However, it was a simple case of economics marching on following the too-often practiced principle of, once the start-up investment has been made, use the resource to extinction and then move on to find another replacement resource.

The great hunt was over when the last herd was killed for food by Sitting Bull's people on the Cannonball River of North Dakota in early 1883. The herds disappeared so abruptly that the white hide hunters bought new outfits for the hunt in 1884, not realizing that the herds were no more. In the end it didn't matter whether one believed that there was another great migration coming, or in the supernatural origin of the bison herds – the great herds were gone.

A Kiowa legend tells of a white buffalo cow that led a small herd of injured, starving and diseased buffalo into a bison paradise to end the era. After 13,000 years the great hunt was over. The buffalo-free plains were left to the large numbers of carrion-fed wolves. Bone collectors now ranged over the grasslands, picking up the skeletons to be shipped east to Michigan and other areas to be converted into phosphate fertilizer.

Much of bison diversity vanished with the great herds. Buffalo of the eastern herds and of the southern plains herd tended to be smaller than the buffalo in the herds of the northern plains and Canada. Other buffalo lived in small groups using the meadows and forests in the Rocky Mountains. These mountain bison were intermediate in characteristics between plains and wood bison. Other differences were noted in the hunt for hides. A gray or 'blue hide' brought $16 as compared to $3.50 for an ordinary hide. Only about one in a hundred hides were of this blue quality. The 'ivory' or 'buckskin' colored hides were even rarer and more valuable. They apparently occurred only once in tens of thousands of hides. The one in 10 million pure white hide was a jewel to both the hide hunters and native Americans. Living white buffalo provided a connection to the supernatural and were of extreme value to the tribes.

By 1885, about 250 wood buffalo remained out of an estimated 170,000 present a century earlier. Only about 1000 plains buffalo remained from their millions. Efforts were begun to save them as the numbers declined. Some of the former buffalo hunters, such as Yellowstone Vic, trapped live animals for zoos and private preserves. A number of ranchers collected buffalo and kept them with their cattle herds. Charles Goodnight of Texas captured five buffalo and built a herd of about 250. Three of his bulls, from the southern plains, were used to add genetic diversity to the remnant herd of about 50 in Yellowstone National Park. Today, the remnant of the Goodnight herd is a group of about 40 animals with dangerously low genetic diversity in a Texas State park.

A plains buffalo displays its surprisingly narrow body.

Walking Coyote (also known as Samuel Wells) of the Salish tribe captured six young buffalo in northern Montana in 1878 and led them several hundred miles over the mountains to the Flathead Valley in western Montana. Two native North American ranchers, Charles Allard and Michael Pablo, purchased them. Additional animals were bought for the herd from ranchers that had a few. The efforts of Pablo and Allard were not just a matter of making money but also a matter of preserving the native American heritage. The herd grew to about 800 under their care. The sale of this herd was forced when the U.S. Congress reduced the size of the Flathead Indian Reservation. The Canadian government bought 700 of the animals in 1907. Others were used to stock the National Bison Range at Moiese, Montana, and 18 were shipped to Yellowstone Park. In 1900 it was estimated that 80 percent of the buffalo remaining in the United States could be traced to the Pablo-Allard herd.

McKay and Bedson acquired five buffalo that had been captured by native Americans in Manitoba. This herd expanded ten-fold. Bedson then sold the herd to Buffalo Jones of Kansas. Buffalo Jones set out with his cowboys to capture buffalo and he built up a herd of 57 over a period of four years. He then added the ones from Manitoba. In 1893 he sold 26 head to Pablo and Allard.

The herd in Custer State Park in South Dakota was started in 1914 with 36 animals from the Scotty Phillips / Pete Dupres ranch herd. Additional buffalo were added from the Pine Ridge Native American Reservation and nearby Wind Cave National Park during the mid 20th century. The herd's size is currently maintained at 900 to 1400 animals.

From the 1790s on there was considerable interest in hybridizing cattle and bison. Many ranchers that had captured buffalo attempted hybridizing. A tame bull mating with a buffalo cow seemed a preferred method of crossing. The offspring, or 'cattalo', are viable but have reduced fertility. Some people regarded the cattalo as being well formed and very promising. Others regarded them as 'bastardly creatures that no one wants.' These efforts led to the presence of a few cattle genes in most of the herds that we have today. More recently hybridizing buffalo with Charolais and Hereford cattle to produce 'beefalo' has been tried. Once again the disadvantages of the hybrids appear to outweigh the advantages.

Conservation

About 18,000 buffalo are in 50 'conservation reserve' populations. Most of these are publicly owned herds of plains buffalo. The free-ranging Yellowstone National Park herd of over 4200 is the largest of these. This herd and the neighboring Jackson Hole herd of over 500 have brucellosis. There are about 4500 wood buffalo in four free-ranging herds in Wood Buffalo National Park, Slave River Lowlands and Peace River areas in Canada. About 4000 wood bison are in 11 other disease-free herds. Eight of these herds are free ranging. All of the existing wood bison herds are in Canada.

The number of privately owned plains buffalo on ranches and native American reservations increased from about 70,000 in the 1970s to over 350,000 in 2002. Privately owned herds are in most Canadian provinces and states of the United States. These populations range in size from small numbers in 'hobby herds' to 4000 or more in commercial production herds.

Disease is regarded as the major obstacle to the establishment of new free-ranging populations. Brucellosis afflicts the Yellowstone National Park herd, which is the largest of the free-ranging plains buffalo conservation herds. Disease is also a problem in four free-ranging wood buffalo herds in Canada. Other conservation obstacles include finding locations for the re-introduction of free-ranging herds, maintaining genetic diversity and providing for situations which allow nature to affect the herds in its time-honored fashion. The situation is also complicated by the commercial raising of buffalo as livestock while also having wild populations. The commercial raising of buffalo has different goals than do the conservation programs, which focus upon maintaining wild free-ranging populations and maintaining genetic and regional diversity. A further complication is that buffalo are legally classified as both livestock and as wildlife. This brings conflict in which agencies have the primary responsibility over them. When the same species of animal is raised commercially in areas where there are wild populations, illegal trapping of wild animals for financial profit may occur.

Bovine tuberculosis is a serious disease of lungs and bone that can also infect cattle and humans. Public health efforts that test for it and then slaughter infected animals have eliminated it from cattle herds, when carried out over a period of years. However, this is a long, complex and expensive process that is difficult to carry out on wild populations. The disease is present in the wood buffalo herds in and around Wood Buffalo National Park. Recent years have brought major expansions of the livestock industry in areas near the park. This has intensified a complex and controversial situation, with many conflicts between commercial interests and those supporting the wild herds.

Brucellosis produces abortion in the first pregnancy in buffalo but it is also a serious disease of both livestock and humans. It also reduces the breeding potential of the bulls that contract the disease. The bacteria produce undulant fever in humans, which antibiotics are not always effective in controlling. Brucellosis is also considered to have potential as a biological warfare agent. Its presence has important public health and international trade consequences for both Canada and the United States. A decades-long control program has eliminated it from cattle herds in both countries by testing and then slaughtering any animals that test positive for it. The bison herds are the major reservoirs of infection remaining in the United States and Canada. The control program has cost several billions of dollars. Cattle are vaccinated against brucellosis when circumstances indicate. The test and slaughter procedure was used to eliminate the disease from buffalo herds in Custer State Park and Wind

Buffalo swing their heads to clear the snow, forming the numerous feeding craters seen here. The fresh snow on their backs demonstrates their excellent insulation which leaves the surface of their coat at air temperature. Note the long hair 'chaps' of the upper front leg of these plains buffalo.

Cave National Park in South Dakota as well as Oklahoma's Wichita Mountain Wildlife Refuge.

Agencies are beginning to use vaccination for brucellosis control in the Yellowstone herd. A system has been developed that uses an air rifle to deliver a 'biobullet' filled with the vaccine into the bison. Vaccines are never 100 percent effective; they were developed for use with cattle, and do not always work well on wild species. The existing vaccines are not as effective on buffalo and elk as they are on cattle. Current vaccines are live organism vaccines, which work by giving an animal a harmless infection to counter a more severe disease. For this reason, predators, scavengers and other animals that might come into contact with the infection produced by the live vaccine must be tested for their response to it before it can be used on wildlife.

The limited size of the National Bison Range in Montana makes it necessary to round up surplus animals.

The Jackson Hole elk herd has a high level of brucellosis infection and uses the same areas as the buffalo herd. This is a severe complication in the attempts to control the disease in the Yellowstone and Jackson Hole buffalo populations. These elk are artificially concentrated on feeding grounds during the winter. This creates an abnormally dense population, which is an ideal situation for the transmission of the disease within the elk herd and also to the Jackson Hole buffalo herd. High real estate prices and other factors involving this popular and expanding resort community make solution of this problem difficult.

The Yellowstone National Park buffalo have also generated major controversy. In spite of the inhibiting effects of brucellosis upon population growth, the herd has grown to an all-time high in numbers. It is currently about seven times larger than the highest population estimate for pre-settlement times. The population has grown large enough that it has begun to migrate out of the Park into neighboring mountains and valleys. This has increased the probability of diseased buffalo coming into contact with cattle, providing a potential for disease transmission.

A buffalo in a squeeze chute where it can be tested for disease.

The efforts of the Montana Department of Livestock to control this movement of buffalo have produced a major controversy, which has brought about research that has greatly increased our knowledge about

Buffalo frequent the geothermal areas of the Yellowstone volcano. The warm ground keeps the snow shallow and plants growing in winter. Buffalo sometimes break through the thin crust and die in the hot springs, or perish when winter temperature inversions trap toxic gases.

brucellosis and its ecology as well as about buffalo ecology.

Genetic concerns include using animals from the most diverse gene pools to start new populations and to use animals that have not been 'contaminated' with cattle genes. Efforts to produce hybrids have introduced some cattle genes into commercial herds as well as some conservation herds. Commercial raisers also select against unruly and older animals and for other characteristics. These efforts will eventually modify the genetic base of these herds and select out 'wild' characteristics. These factors focus conservation efforts on the more genetically diverse publicly owned conservation herds. There are a few private herds that have been designated conservation herds because their owners have agreed to manage their herds for genetic conservation.

The Yellowstone herd has the most robust plains bison gene pool. Quarantine facilities are being developed to allow disease-free animals from Yellowstone to be used to start other populations. Buffalo calves from the Yellowstone herd that test negative for brucellosis will be held in the quarantine facility. This is being done to eliminate the problem of latent infections that show up after a period of time even in animals that test negative for the disease.

Bison conservators wish to conserve the genetic pools of the two subspecies separately rather than letting them interbreed, as has sometimes been proposed. Beginning in 1963, Canada reestablished disease-free wood buffalo herds in British Columbia, Manitoba, Northwest Territory and the Yukon. The Mackenzie River herd of 2000 animals is the largest of these herds. Alaska has expressed interest in establishing wood bison populations in addition to its four plains bison herds. Sergey Zimov has proposed transplanting Canadian wood buffalo to Siberia as part of the 'Pleistocene Park' project. This is an experiment to use horses, bison and other survivors from the ice age to recreate the grasslands present during the time of the ice. The goal of recreating the 'mammoth steppe' or grassland is to reduce the massive release of carbon dioxide and methane expected from the Arctic as global warming occurs.

No specific areas have been formally proposed for the reintroduction of wild free-ranging plains bison herds, although the Missouri Breaks area of eastern Montana has often been mentioned. The human population trend is currently towards depopulation of the Great Plains. Because of this, the concept of establishing a 'buffalo commons' is sometimes proposed. But even if human depopulation is occurring, most of the land of the Great Plains is in private ownership and is being used for agriculture or some other economic purpose.

As buffalo populations have grown and reached the limits of their habitats, the 13,000-year-old hunting relationship to man has been revived. Buffalo are currently hunted in Alaska, Arizona, British Columbia, Montana, Northwest Territories, South Dakota, Utah, Wyoming and Yukon. They are also hunted on some private ranches and native American reservations. Hunting tends to maintain the wariness, adult aggressiveness and other characteristics that buffalo have developed over the centuries in their interactions with humans.

The status of buffalo has improved greatly during the last 100 years, but major problems remain in conserving them for the indefinite future. The job is not done but wildlife biologists, managers and private citizens are very involved in conserving this icon of North America.

The Yellowstone forest fires of 1988 killed many trees, but had little effect on the buffalo. The robust size and genetics of this herd make it a potential resource for establishing new wild herds, but brucellosis sadly prevents this.

Historic Distribution of the Buffalo

PROBABLE DISTRIBUTION
ABOUT 30,000 YEARS BEFORE PRESENT

PROBABLE DISTRIBUTION
ABOUT 20,000 YEARS BEFORE PRESENT

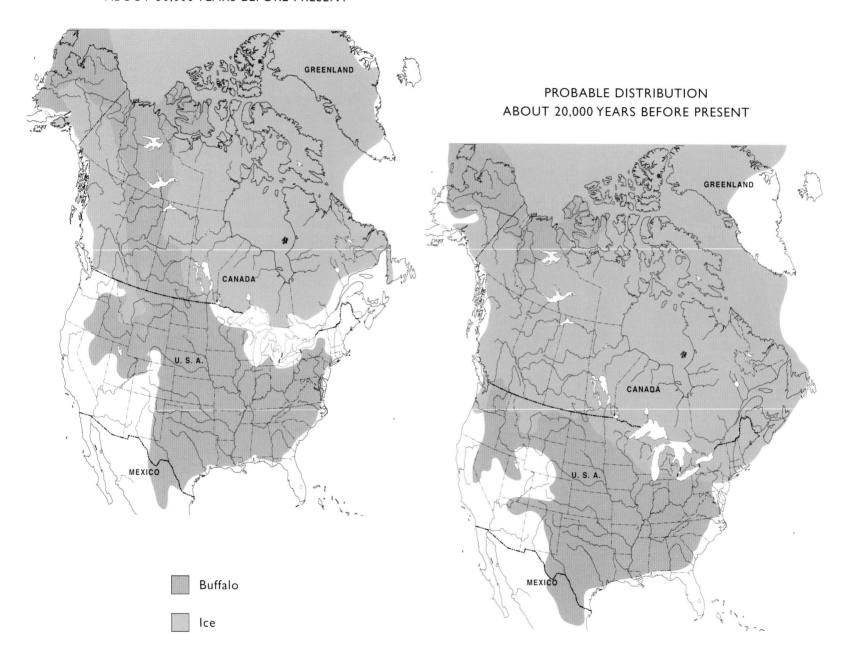

Buffalo

Ice

Historic Distribution of the Buffalo

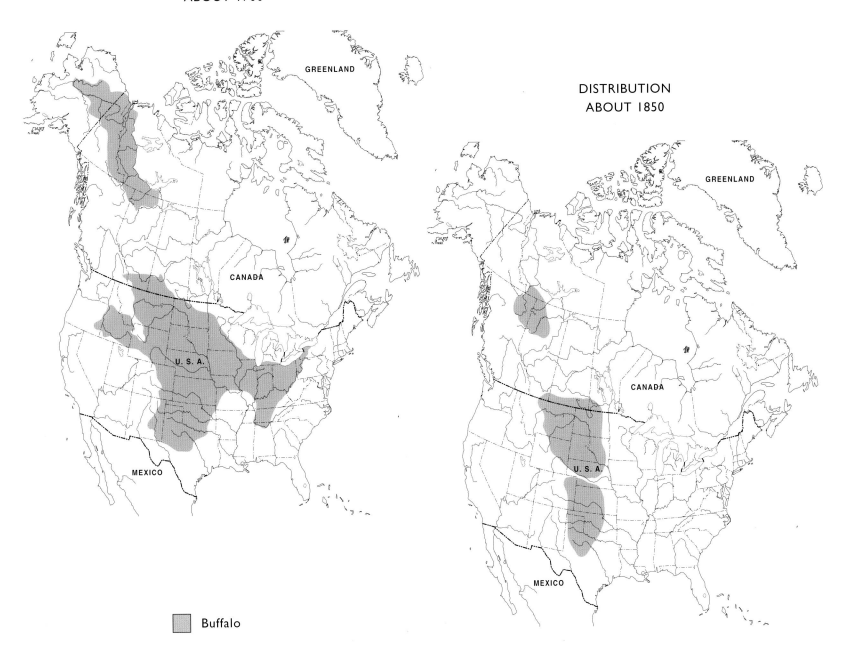

PROBABLE DISTRIBUTION
ABOUT 1700

DISTRIBUTION
ABOUT 1850

Buffalo

Buffalo Facts

Scientific names: *Bison bison* or *Bos bison*
Common names: Bison, Buffalo

		Bison bison bison: **the plains buffalo**	*Bison bison athabascae*: **the wood buffalo**
Subspecies:			
Weight:	Males	1692 lb (769 kg) max. at 13 years	2002 lb (910 kg) max. at 13 years
	Females	999 lb (454 kg) max. at 10 years	1247 lb (567 kg) max. at 12 years
Total Length:	Males	120 to 150 in (304 to 380 cm)	
	Females	83 to 125 in (210 to 318 cm)	
Shoulder Height:	Males	66 to 73 in (167 to 186 cm)	
	Females	60 to 62 in (152 to 157 cm)	

Gestation Period: 270 to 300 days

Number of Young: Usually 1; twins appear to occur in less than 1 out of 1000 pregnancies. Calves weigh 31 to 40 lb (14 to 18 kg) at birth.

Longevity: Buffalo have lived up to 41 years in captivity. Animals over 15 years are considered old in wild populations.

Age of Sexual Maturity: While cows are known to conceive as yearlings, giving birth to a calf when they are 2, it is usual for them to breed initially at 2, giving birth at age 3.

Speed of Locomotion: They attain speeds up to 36 mph (60 kmph).

Home range size: Varies by age, sex and forage availability. Annual home range sizes for adult wild females range from 154 to 480 sq. miles (398 to 1240 sq. km). Adult males 66 sq. miles (170 sq. km).

Recommended Reading

The Buffalo. The Story of American Bison and Their Hunters from Prehistoric Times to the Present. Francis Haines. University of Oklahoma Press, 1994.

I will be Meat for my Salish R Bigert (Ed.) Salish Kootenai College, Pablo, MT, 2001. A book of Native American accounts by The Montana Writers Project and the buffalo of the Flathead Indian Reservation.

'Bison'. Chapter 48 in *The Wild Mammals of North America*, by H Reynolds, C Gates and R Glaholt, Johns Hopkins University Press, Baltimore, 2003. This is a detailed 51-page summary of the scientific knowledge about wild bison.

Going to the Buffalo: Indian Hunting Migrations across the Rocky Mountains by W Farr, in *Montana: The Magazine of Western History*, Winter 2003 and Spring 2004. A two-part authoritative historical article concerning the hunting of the northern plains bison herd.

Frozen Fauna of the Mammoth Steppe: The Story of Blue Babe. R D Guthrie, University of Chicago Press, 1990. This book tells the story of the frozen fossil bison and discusses the evolution and ecology of modern bison.

Biographical Note

Harold Picton received his B.S. in Fish and Wildlife Management from Montana State University in 1954, a M.S. in Fish and Wildlife Management from Montana State University in 1959 and a Ph.D. in Physiology from Northwestern University in 1964. He worked in the Zoology and Entomology Department from 1963, and became a full professor in 1976. Dr Picton joined the Fish and Wildlife Management Program in the Biology Department in 1976. Latterly he specialized in teaching Animal Physiology and courses in Fish and Wildlife Management.

Dr Picton has co-authored many articles on Animal Physiology and Fish and Wildlife Management, including papers in the prestigious international journal, *Nature*. His monograph on the history and management of the Sun River game range, *Saga of the Sun*, stands as a benchmark for similar publications. He collaborated with colleagues in Montana Fish, Wildlife and Parks, the U.S. Forest Service and the National Park Service on many cooperative wildlife projects. Dr Picton was honored by the Montana chapter of the Wildlife Society with the Distinguished Service Award for this outstanding work in the field of wildlife biology and management.

Harold Picton is Professor Emeritus of Fish and Wildlife Management at Montana State University.

Index